ECHOES MAGAZINE

VOICES

Series 1

May 2020

EDITOR: JANE LANDEY

CONTRIBUTING EDITOR: JONATHAN EMMANUEL

ISBN: 9798648667549

PUBLISHED BY BRIMS WORLD

PRINTED IN U.S.A.

PREFACE

"Knowledge is power", so the adage of the wise ones. Echoes magazine is published twice a year and copies will be available to satisfy your curiosity. Look out for copies in May and October of every year! In this first edition, our special readers will experience thorough knowledge on different aspects of life. Reading about people and their origins which include archeology, history, geology etc. Extracts and written facts will reflect in the presentations. This is taking you back to foundation of events and process of growth. General acquisition of knowledge widens your understanding and do not limit you to a specific area. Therefore go for more!

Jane Landey

The editor

Medical Science

How do you know the condition of your body? There are facts about life that have been in vogue from time immemorial. One fact remains, health is wealth and nothing can replace it. Coping with your health is very important and remaining healthy is far more important.

COPING WITH YOUR HEALTH

There are several ways you can accommodate good health. The following are clues to doing this:

1) Eat balanced diet

Food is the friend of a good complexion. Healthy body comes from a good diet! A balanced diet comprises of the following:

Proteins, carbohydrates, fats and oil, vitamins, minerals, and water.

Proteins

Carbohydrates

Fats and oil

Vitamins

Minerals

2) Drink clean water.

Water

3) Live in a clean environment

4) Visit a physician when sick

When you experience any symptom of illness, do not prescribe medication for yourself.

Arrange to visit a doctor that can diagnose the ailment you have. He will tell you what you must do to become fit and healthy again.

This is proper because buying drugs illegally to treat yourself will worsen your health!

Technical Way Round the Home

The world is going technical and it is actually in favor of producers. Equipment is used this day to put more in production. I can remember when I was young cutlass and hoe were used to cut down tall grasses, a clean job was rendered. If the two tools were not enough to do it, both hands were added to finish it. Nobody thought of using gloves but both hands were thoroughly washed!

Today, little or mass work is done to remove grasses from your yard, frontage or football field. What do people use presently to remove grasses of any level? This brings us to new equipment such as mower, grass cutter and scythe.

The following are types of machines and tools for removing weeds and tall grasses. Lawn mower, brush mower, trimmer, edger and blowers. Basically, it is ideal to cut the grasses in your yard or lawn. This allows you to see any animal crawling by or hiding around your home. Technology has brought fair play to all areas of business which makes it easier for mankind. Some parts of the world are yet to know the full impact of modern equipment simply because the people cannot afford it or they do not know it exists. There is a great advantage to people of the nations where they are produced.

THE USES OF GRASS CUTTING TOOLS

Lawn mower

A lawn mower is a machine that cuts a grass surface to equal level. This machine is ideal for reducing the grasses on your land to the level you desire.

Brush mower

This equipment can cut through grass or weeds three inches thick and six inches tall. You have to hold the break on one side and it will make its way through.

String Trimmer

A trimmer is a handy yard tool that enables you to cut grasses and weeds growing in places your mower cannot reach.

Lawn edger

This tool is used to mark out boundaries between a lawn that has grass and another ground surface.

Leaf blower

A leaf blower is a tool used in the garden to blow leaves away from the ground. The manual way of removing leaves from the ground is by using a rake. A leaf blower reduces the energy one will have to use. It is much easier than a rake.

All is set for you and just get to work immediately. Put on old clothes or work clothes if you have one. The dirt and dust will fall on the clothes you are wearing and you will not need to worry if they are to be discarded!

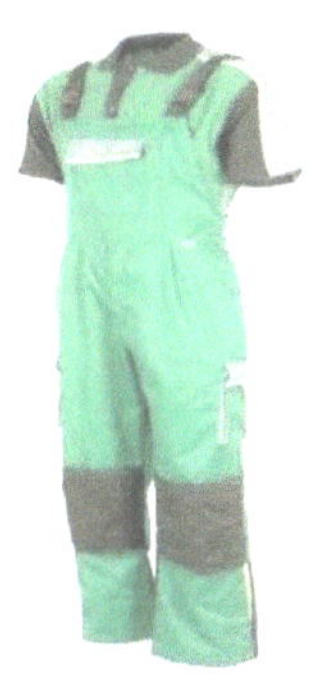

Gloves are very important, this will cover your hands and they will not get stained or dirty. Thick garden gloves are better than ordinary gloves.

There are boots fit for working in the garden or yard. These boots can trample on stumps and broken branches without hurting your feet.

The weather could be sunny and hot therefore carry along your round hat that covers both your head and face.

THE WORLD OF ENTERTAINMENT

Medieval period takes us back to the fifth century, also known as the Middle ages. People in some part of the world such as Western Europe had to depend on sacred vocal music. Latin had to be sung in churches which was simple and pure. The spread of Christianity at the time paved way for popularity of hymns and secular songs. Thus the two types of music during the medieval period were sacred and secular music. The sacred music was religious while the secular music was non-religious. The medieval instruments are, flute, harp, lute, mandolin, recorder, bagpipe, bladder pipe, tambourine, finger cymbals, shofar, serpent, psaltery and gemshorn.

The father of medieval era.

St. Augustine of Hippo
A theologian and a devotional writer.

Renaissance era spread rapidly between fifteenth and sixteenth centuries. It is also called a period of rebirth in arts, science and philosophy. It was a period of transition from the ancient world to the modern. It was quite different from the medieval era, whereby music was important in Catholic and Protestant churches. Music also was important for entertainment in courts and dance. Renaissance listened to music from different organs such as psaltery, harp, shawm, bagpipe, drums and flute.

The father of renaissance.

Francesco Petrarca

Humanist philosopher

Baroque period dated back to the seventeenth century and became widely known in the nineteenth century in Western Europe for the art music. This was an elegant style of art represented by curving lines, gilt and gold. The instruments used in Baroque music were strings. woodwind, brasses, keyboards and percussion. Baroque contributed to the usage of instruments for entertainment. It was in this period that musical genres such as opera, concerto, cantata, sonata and oratorio were in vogue.

The father of baroque era

Michelangelo Merisi Caravaggio

Painter

Classical era, also known as Classicism was a period of classical artistic and literary heritage of Greece and Rome. It was the first era in music history when public concerts were highly involved. It began in the mid seventeenth century and ended in the early nineteenth century. The music of this era was more of simplicity and had the following features, elegance and balance, display of contrasting moods, simple diatonic harmony with melody and accompaniment. The instruments used were, strings, woodwinds, keyboards, brasses, and percussion.

The father of the symphony and father of the String Quartet

Franz Joseph Haydn

Romantic era began in Europe at the end of the eighteenth century and lasted by the mid nineteenth century. It emerged from the classical period and had great influence on arts, literature and music of the Europeans. The Romantic poets became a part of the movement that wanted liberty and an end to the oppression of the poor. There were different notable instruments used in music. The following instruments featured prominently during this period: strings, woodwind, brass, percussion and key.

Strings

Woodwind- flute and piccolo   oboes and clarinets   bassoon and double bassoons

Brass- trumpets   trombones   French horns   tuba

Percussion-

Key- piano

Father of Romanticism

Jean-Jacques Rousseau

Earth Science

Ecology fills us with the ideas of what our rocks have in store for us. Rocks decay over many years to accumulate substantial minerals we take advantage of this present day. Our geologists go on days and months working on sites that could lead them to iron ores, petroleum coal etc. Geology gives insight into the structure of the Earth above and below. It helps us to know the history and ages of rocks in specific locations by using certain tools.

Geology is important for exploring minerals and hydrocarbon. It further helps us to understand natural hazards and solutions to environmental problems. We are also able to access water resources and delve into past climate change.

These branches of geology have shaped the lives of mankind in different ways.

Structural geology

This is the study of the evolution of a particular area which has to do with the widespread patterns of rock and deformation. There are mountain building, land rifting and other formation due to plate motion.

Mining geology

The method to extract mineral resources from the earth surface is called mining. Some of this resources are for economic purposes and they serve humans needs. They include, copper, gold, metals, asbestos, phosphates, clay, silica etc. Sulfur, helium and chlorine are elements that could be extracted from beneath the earth if found.

Petroleum geology

Petrol and natural gas are extracted from the earth which had been in sedimentary basins. The formation of these basins is studied in regards to their evolution and present forms.

Petroleum is a yellowish –black liquid extracted from beneath the earth surface. It is formed when large quantities of dead organisms such as algae and zooplanktons are buried under sedimentary rock and are exposed to intense heat and pressure. It is refined and distilled into numerous consumer products.

Usage of Petroleum

Gasoline

Kerosene

Asphalt

Chemical reagents

Plastics

Pesticides

Pharmaceuticals etc

Petroleum is also referred to as crude oil.

Natural gas comes from the decomposition of plants and animal matter that have been exposed to heat and pressure under the surface of the Earth millions of years back.

Usage of Natural gas

Heat

Cooking

Electric generation

Fuel for vehicles

Organic chemicals

 Natural gas is simply referred to as gas.

Engineering geology

This field of geology is the application of geological factors such as construction, design, location, operation and maintenance of engineering works.

Economic geology - It is concerned with earth materials that can be used for economic and industrial purposes.

Planting Season

It is time to plant your vegetables and crops. What do you have on mind this time around?

Lettuce, garden eggs, watermelon, cucumbers, carrots, spinach, tomatoes, pepper etc.

Vegetables grow fast when properly cared for. The season of growth also matters and the soil must be rich enough for planting.

Let us talk about the season good enough for planting.

We need rainfall and temperature to grow plants. Plants have certain seasons for adaptation to grow. Planting starts at the beginning of raining season or spring which varies in different parts of the continents. People start making rows on their garden in order to plant edible plants such as vegetables and common crops. They plant spinach, cabbage, onions, carrots, beets, lettuce, asparagus, chilies, potatoes, beans, peas, etc. The best vegetables to plant in March in some parts of the world include, spinach, beets, beans, carrots, lettuce, cauliflower, onions, pas, radishes, brussels sprouts, turnips etc.

How do you start your early planting?

The following are tips for you.

1) Buy quality seeds, seedlings and fertilizers
2) Get your garden ready by cutting off all the grasses. Clear them and park them aside.
3) Get the soil ready to plant your seeds.
4) Plant the seeds with enough space for each.
5) Water your plants regularly and apply fertilizers often.

Seeds

Tomato seedling

Cultivated soil

Planting the seedlings

Removing weeds

Lettuce growing

What to plant in May

Cabbage

Squash

Fennel

Pea

Melon

Sweet corn

Computer Literacy

Many people are conversant with computer operation in recent time. They are either forced to learn it because they must be computer literate to get a well paid job. Once they master using the computer, the rest is simple; they become competent as they regularly operate it. The computer has all the details of how one can operate it because as soon as you open it up, it tells you what to do even if you do not know it.

Types of Computers

Personal computers

Desktop, laptop

Personal computers have screens that pop up words and tell you what to do. They do numerous things such as writing, editing, communication etc.

Embedded Computers

Mobile phones, automatic teller machines, microwaves ovens, cars, alarm clocks and CD players.

They are embedded computers because they do one thing only and do it well.

Here are some home uses of computers:

Writing

Watching videos

Listening to music

Playing computer games

Using the internet

Some uses of computers at work:

Excel

Word processing

Software development

Recording

PowerPoint

Internet

Editing

Computer Program

This describes the instructions stored as a file on the computer's hard drive. When a user runs the program, the file is read by the computer. The processor reads the file as a list of instructions which makes the computer performs the task.

Here are some examples of computer programs

Operating system

Spreadsheets

Office suite

Video games

Web browser

Computer analysis can help you get things done fast. If you need to write a book or your memo, computer helps you do this and corrects your mistakes by showing red underline. If the words are not properly embedded, it shows green underline. You can send messages through your e-mail to a friend, family of coworker as fast as you can. You can also amuse yourself with music or movies.

What else could be more real and fun!

Medical

Technology

Agronomy

Ecology

Archeology